Evolution — Fact or Fiction?

John Blanchard

EVANGELICAL PRESS
Faverdale North, Darlington, DL3 0PH, England

e-mail: sales@evangelicalpress.org

Evangelical Press USA
P. O. Box 825, Webster, New York 14580, USA

e-mail: usa.sales@evangelicalpress.org

web: http://www.evangelicalpress.org

© Evangelical Press 2002. All rights reserved. No part of this publication may be reproduced, stored in a retrieval system or transmitted, in any form, or by any means, electronic, mechanical, photocopying, recording or otherwise, without the prior permission of the publishers.

First published 2002
Second impression January 2003
Third impression September 2003
Fourth impression August 2004
Fifth impression January 2006
Sixth impression March 2009

British Library Cataloguing in Publication Data available

ISBN 13 978-0-85234-530-6   ISBN 0-85234-530-5

Scripture quotations in this publication are from the Holy Bible, New International Version. Copyright © 1973, 1978, 1984, International Bible Society. Used by permission of Hodder & Stoughton, a member of the Hodder Headline Group. All rights reserved.

Printed and bound in Great Britain by Castleprint, Richmond, Yorks.

Evolution — Fact or fiction?

The picturesque English village of Downe lies about fifteen miles southeast of London, but the two communities have always been light years apart in virtually every way. In the middle of the nineteenth century, London was the largest trading centre in the world, had a population of three million, and was a focal point for many of the latest advances in science and technology. In Downe, home to fewer than 500 people, village life meandered gently along, centred on the old flint parish church and the local infants' school. Contemporary records listed the leading citizens as including the butcher, the baker (and postmaster), the grocer and the boot and shoe maker. Asked which of these two places would host an event that would transform man's thinking about his place in the world, people would surely have chosen London. They would have been wrong!

While most of the villagers gently eased through each day as it came, the wealthy owner of Down House, a quarter of a mile from the village centre (and oddly spelled without the 'e') was engaged on a project that consumed him day and night. A doctor's son from Shropshire, he had begun to study medicine at Edinburgh University. But when he felt unable to face the prospect of a lifetime dealing with gruesome medical problems, he switched to reading classics, mathematics and theology at Cambridge, with a view to a career in the Church of England ministry. He soon abandoned this idea, yet ironically his theology results made up for poor marks in classics and even worse ones in mathematics and earned him a B.A. degree in 1831.

His health was often indifferent and by the time he went to live in Downe he was something of a hypochondriac, keeping extensive notes on his numerous symptoms, medicines and treatments. Yet his illness gave him a useful excuse to avoid getting involved in the social life of the village. He was virtually a recluse, spending almost his entire time pursuing his project. In the event, he never completed the massive work he intended to publish, but in November 1859 a condensed version of 490 pages was rushed into print. Its title — *On the Origin of Species by Means of Natural Selection or the Preservation of Favoured Races in the Struggle for Life* — was hardly snappy or attention-grabbing, yet the book took the world by storm. Nearly 150 years later its main thesis is still being debated with more passion than virtually any other subject in science. It goes under the general name of 'evolution'; the author of the book was Charles Darwin; and his influence is now so far-reaching that in the course of the BBC's 'Evolution Week' in 1998 he was described as 'the man who killed God'.

Definitions

Strangely, the word 'evolution' is nowhere to be found in the original version of Darwin's book (now mercifully known by the abbreviated title *The Origin of Species*, or even *The Origin*) and 'evolved' did not appear until the sixth edition, but the *Concise Oxford Dictionary* puts its finger on what Darwin had in mind: 'origination of species by development from earlier forms, not by special creation.'[1] The last four words in that definition are the catalyst for today's debate on the subject. On one side are those who believe that by natural processes all living things have arisen from a single common ancestor, which in turn had an inanimate source. On the other side are those who believe that a transcendent Creator, using processes that are not in operation today, brought the many different life forms into separate existence.

Evolution — Fact or fiction?

Yet before we go any further, we need to make a distinction between two forms of evolution. Seven chapters of *The Origin* were taken up with *micro-evolution*, the theory that in organisms *of the same kind* different characteristics emerge as the result of adaptation to their particular environment. Although Darwin called this his 'special theory', there was nothing new or controversial about it. The development of different varieties within species had already been widely accepted, and in the year that Darwin was born the French botanist and zoologist Jean Baptiste de Lamarck anticipated a great deal of what he wrote. Today, nobody seriously denies that perfectly natural processes produce gradual changes within species, and even new species; or that over a long period of time these can result in considerable variations in the size, colour and other characteristics of plants and animals. We can even accelerate such changes by deliberate crossbreeding or genetic engineering.

If Darwin had stopped at micro-evolution, his name would by now be virtually unknown. What propelled him into the headlines — and has kept him there ever since — was his 'general theory' of *macro-evolution.* This revolutionary idea claimed that natural processes produce *new species without limitation,* and that all species can be traced back to a single common ancestor. Twelve years after the publication of *The Origin,* Darwin went a gigantic step further and claimed that *man* was included in this universal evolutionary process. In *The Descent of Man* he wrote, 'The main conclusion here arrived at ... is that *man is descended from some less highly organized form.*'[2] To make it clear that by then he saw no place for God in his 'molecules-to-man' scenario, he added that humankind had survived '*not according to some ordered plan* but as the result of chance operating among the countless creatures by nature's unlimited tendency towards variation'.[3]

In Darwin's model a key role was played by what he called 'natural selection'. This said that populations of organisms develop new characteristics in response to 'selective pressures' in their environment (more simply, in order to survive) and that when these new characteristics become permanent, new species emerge. In the sixth edition of *The Origin,* Darwin adapted a phrase coined by the eccentric British philosopher

Herbert Spencer, and said that 'the survival of the fittest' was a 'more accurate' explanation of what he meant by 'natural selection'.

Scope and support

Darwin's 'general theory' says that all living species, from ants to antelopes, broccoli to bats, carrots to cockatoos — and hummingbirds to humans — are the result not of imaginative design, but of chance variations, with all living things traced back to an original spark of life that appeared on our planet at some point in prehistory. This is such an amazing scenario that even Darwin had 'awful misgivings' about it. He reworked so much of the manuscript of *The Origin* that he offered to pay the publisher for the cost involved. It has been pointed out that in *The Origin* and *The Descent of Man*, phrases like 'we may suppose' occur over 800 times, yet within about twenty years his highly speculative hypothesis had gained such a foothold that the outrageous had become orthodox.

Today evolution goes far beyond biology, and claims to explain the origin, arrangement and development of everything in the universe without any need for a transcendent Creator. As geneticist Theodosius Dobzhansky puts it, 'Evolution comprises all the stages of the development of the universe: the cosmic, biological, and human or cultural developments.'[4] The British biologist Sir Julian Huxley called it 'the most powerful and the most comprehensive idea that has ever arisen on earth',[5] and the Australian molecular biologist Michael Denton is not exaggerating when he writes, 'The idea has come to touch every aspect of modern thought; and no other theory in recent times has done more to mould the way we view ourselves and our relationship to the world around us.'[6] The evolution lobby has got its message across so effectively that any student in a school, college or university who challenges it is likely to be written off as an eccentric, but this is hardly surprising when it has such influential backers.

- In 1966 the celebrated geneticist H. J. Muller circulated a manifesto signed by 177 American biologists asserting that the organic evolution of all living things, man included, from primitive life-forms, and even eventually from non-living materials, is a fact of science as well established as the fact that the earth is round.[7]
- Astronomer Carl Sagan, best known for his ground-breaking television series *Cosmos*, stated in the printed version of the programmes, 'Evolution is a *fact*, not a theory.'[8]
- The Oxford zoologist Richard Dawkins, Simonyi Professor of Public Understanding of Science (and arguably Britain's best-known atheist), makes the point with his usual panache: 'It is absolutely safe to say that, if you meet somebody who claims not to believe in evolution, that person is ignorant, stupid or insane (or wicked, but I'd rather not consider that).'[9]

With heavy-hitters like these endorsing it, the theory of evolution is enthusiastically echoed in the media. Virtually all nature programmes on radio and television take it for granted, and in discussions on the subject those who take the opposite view are often given no more than token parts. With most biology textbooks adding their *imprimatur*, the case for evolution seems cut and dried. Yet for Sir Julian Huxley to claim that 'all scientists agree' with evolution, and that '... there is absolutely no disagreement'[10] is simply not true.

- Ambrose Fleming, one-time president of the British Association for the Advancement of Science, called evolution 'baseless and quite incredible'.[11]
- Nobel Prize winner Robert Millikan states, 'The pathetic thing is that we have scientists who are trying to prove evolution, which no scientist can prove.'[12]
- Colin Patterson, Senior Palaeontologist at the British Museum of Natural History, agrees: 'It is easy enough to make up stories of how one form gave rise to another ... but such stories are not part of science.'[13]

These views are corroborated by the American engineer Henry Morris after an exhaustive study of *The Origin:* 'One can search the whole book in vain for any real scientific evidences of evolution — evidences that have been empirically verified and have stood the test of time. No proof is given anywhere — no examples are cited of new species known to have been produced by natural selection, no transitional forms are shown, no evolutionary mechanisms are documented... One can only marvel that such a book could have had so profound an influence on the subsequent history of human life and thought.'[14]

Fossils: fiction and the facts

Those who promote the theory of evolution employ data from cosmology, physics, biology and genetics, but when it comes to discovering what happened in the past the big picture is palaeontology, the study of fossils. This is hardly surprising. If life arose, developed and became increasingly more complex through a random evolutionary process, fossils should indicate that there were countless transitional stages between the different species. The distinguished French zoologist Pierre Grassé stated, 'Naturalists must remember that the process of evolution is revealed *only through fossil forms* ... only palaeontology can provide them with the evidence of evolution and reveal its course or mechanisms.'[15] In 1981 a spokesman for the American Association for the Advancement of Science claimed that 100 million fossils, identified and dated, 'constitute 100 million facts that prove evolution beyond any doubt whatsoever'.[16]

To Darwin's embarrassment and frustration, the fossils did no such thing. He expected an 'inconceivably great' number of intermediate and transitional forms, but geologists discovered species, and groups of species, that appeared to have neither predecessors nor successors. Darwin soon realized that the absence of any finely graduated organic chain was 'the most obvious and serious objection against the theory [of evolution]',[17] and his response was to suggest that later discoveries would fill the gaps. Yet as evolutionist David Raup, Curator of Geology at Chicago's Field Museum of Natural History, has pointed out, this has not happened: 'The evidence we find in the geologic record is not nearly as compatible with Darwinian natural selection as we would like it to be... We now have a quarter of a million fossil species but the situation hasn't changed much. The record of evolution is surprisingly jerky and, ironically, *we have even fewer examples of evolutionary transition than we had in Darwin's time ... so Darwin's problem has not been alleviated.*'[18] Forty years of research led Professor N. Heribert Nilsson, of Lund University, Sweden, to write, 'It is not even possible to make a caricature of evolution out of palaeobiological facts. The fossil material is now so complete that the lack of transitional species cannot be explained by the scarcity of the material. *The deficiencies are real; they will never be filled.*'[19]

Some evolutionists have tried to wriggle out of this by inventing the idea of *punctuated equilibrium*, which suggests that millions of more or less static years (equilibrium) were occasionally interrupted by worldwide cataclysms (punctuations) that resulted in wholesale extinctions and made way for radically new life-forms. This seems like a promising escape route but it turns out to be an argument from silence, as there is not a shred of evidence to back it up. As the American scholar Marvin Lubenow wryly comments, 'It must be the only theory put forth in the history of science which claims to be scientific but then explains why evidence for it cannot be found.'[20] Others have suggested the so-called 'hopeful monster' theory, which includes the idea that the first bird was hatched from an egg laid by a reptile. This would be a novel arrangement, but would have had to happen at least twice at the same time and in the same place; otherwise the first bird would have had no mate and therefore no offspring. We should not confuse science with science fiction! When the man in the street sees impressive collections of bones arranged in a certain way, he tends to take it for granted that they are part of the evidence for evolution, but David Kitts, Professor of Geology at the University of Oklahoma — a convinced evolutionist — was honest enough to put the record straight: 'Evolution requires intermediate forms between species *and palaeontology does not provide them.*'[21]

Far from backing evolution, the fossil record helps to bury it. The earth's Cambrian Period is said to have begun some 600 million years ago and to have lasted about eighty million years. The Cambrian rocks reveal a bewildering variety of fossils of highly developed life forms, including those of trilobites, sponges, brachiopods, worms, jellyfish, sea urchins, sea cucumbers, swimming crustaceans, sea lilies and other complex invertebrates. Their appearance is commonly called the 'Cambrian Explosion' — and they deal two damaging blows to evolution.

The first is that the Cambrian fossils represent nearly every major group of organisms living today. The 'explosion' is said to have taken less than five million years — a blink of time in evolutionary terms — yet is also said to have initiated virtually all the major forms of multicellular life. As physicist Cornelius G. Hunter shows, this would mean that 'In a geological moment, the fossil species went from small worm-like creatures and the like to a tremendous diversity of complex life forms, including virtually all of today's modern designs.'[22]

The second blow to evolution is that scientists have been unable to trace any sign of precursor life forms in earlier layers of the earth's crust. Even Richard Dawkins admits that this points to the Cambrian fossils being those of fully formed creatures with no ancestors from which they evolved: 'It is as though they were just planted there, without any evolutionary history.'[23] Try as they might, evolutionists have never been able to make the facts fit their theory, and Berkeley law professor Phillip Johnson may be right to call the Cambrian explosion 'the single greatest problem which the fossil record poses for Darwinianism'.[24]

But what about those impressive 'evolutionary trees', festooned with technical names and said to show how man evolved from all kinds of exotic ancestors? To the layman, they may seem reasonable and even convincing, but Colin Patterson, surrounded by one of the world's greatest collections of fossils, says they are neither: 'We have access to the tips of the tree; the tree itself is a theory, and the people who pretend to know about the tree and to describe what went on it — how the branches came off and the twigs came off — are, I think, telling stories.'[25] Stephen Jay Gould, the Harvard Professor of Biology and Palaeontology, confirms Patterson's assessment almost word for word: 'The evolutionary trees that adorn our textbooks have data only at the tips and nodes of their branches; the rest is inference, however reasonable, *not the evidence of fossils.'*[26] This is a devastating indictment! These men (both evolutionists) are saying that the only items of which we can be certain are 'the tips of the trees' — in other words, the separate species themselves — and that, for all the splendid names and impressive graphics, the whole of the rest is an elaborate invention. As Patterson says elsewhere, 'Fossils may tell us many things, but one thing they can never disclose is whether they were ancestors of anything else.'[27]

If the standard evolutionary model is true, we should expect (just as Darwin did) to find the earth teeming with evidence of intermediate life forms — but they are simply not there. If, on the other hand, God created fully formed and separate kinds, we should expect to find the remains of highly complex, fully formed specimens, all without any apparent ancestors — *and that is exactly what we do find*. The fossils have a significant contribution to make in the debate about the origin of species — but they are not saying what evolutionists want to hear!

The campaign to persuade people that the fossil record supports Darwinian evolution has been an almost unqualified success, yet ironically many highly qualified evolutionists say exactly the opposite. Jeremy Rifkin writes, 'What the record shows is nearly a century of fudging and finagling by scientists to conform with Darwin's notions, all to no avail. Today the millions of fossils stand as very visible, ever-present reminders of the paltriness of the arguments and the overall shabbiness of the theory that marches under the banner of evolution.'[28] Stephen Jay Gould openly admits that the lack of intermediate life forms 'persists as the trade secret of palaeontology',[29] while the American Museum of Natural History's Niles Eldridge goes even further and openly confesses, 'We palaeontologists have said that the history of life supports … [the story of gradual adaptive change] … *all the while knowing that it does not.*'[30] These are startling admissions. Here are articulate evolutionists blithely telling us that, with regard to the claim that fossils point to evolution, we have had the wool pulled firmly over our *eyes!*

Then why should intelligent experts cling to the fossils in this way? The simple answer lies in their world-view, the starting-point for all of their thinking and speculation. Prior to Darwin, science was largely spearheaded by those who believed that God created separate kinds. This is no longer the case. Although countless scientists today hold to divine creation, Colin Patterson is right to claim that in many cases, '… post-Darwinian biology is being carried out by people whose faith is in, almost, *the deity of Darwin.*'[31] If God is a non-starter, evolution seems to be the only show in town, and all scientific data must somehow be made to support it, even if the usual rules of assessment have to be suspended. Harvard geneticist Richard Lewontin is perfectly frank about this: 'We take the side of science in spite of the patent absurdity of some of its constructs … *because we have a prior commitment to materialism…* Moreover, that materialism is absolute *for we cannot allow a Divine foot in the door.*'[32] We shall return to this point later, but we dare not miss what Lewontin is saying here, which is to admit that *his science is being driven by his faith in materialism*. The University of Chicago's L. T. More makes the same connection: 'Our faith in evolution depends on *our reluctance to accept the antagonistic doctrine of special creation.*'[33] People who have been hoodwinked into believing that fossils explain away the idea that God created fully formed separate species ought to take careful note of this.

Molecules to man?

If we assume the existence of the first life form — we shall look at this shortly — the overall evolutionary scenario then has protozoans (microscopic single-cell organisms) leading to the first invertebrates (life forms with no spinal cord or backbone); invertebrates evolving into fish; fish into amphibians; amphibians into reptiles; reptiles into birds and furry quadrupeds; and furry quadrupeds into apelike mammals. Yet all of this is pure speculation, with no solid evidence to back it up — and the same can be said about the 'final chapter' in the evolutionary story, in which apelike mammals evolve into *Homo sapiens,* the human race. Evolutionists have pulled out all the stops in trying to establish this particular link, but their best-known efforts have ended in spectacular failure.

- Unearthed in Germany in 1857, *Neanderthal Man* was touted as the vital 'missing link', but the evidence against it has become so overwhelming that most experts now agree with the verdict that 'Neanderthal was a card-carrying member of the human family.'[34]
- For forty years, a collection of bones found in the south of England in 1912, and dubbed *Piltdown Man*, was trumpeted as 'the sensational missing link'.[35] Everybody now knows that it was a gigantic hoax, a 400-year-old human skull married to an orang-utan's jaw.
- In 1922, a single tooth unearthed in Nebraska and said to be up to 5.5 million years old attracted international headlines as proof of a link between apes and humans, but within six years it was found to have come from a peccary, a piglike wild animal that, like the *Nebraska Man* theory, is now extinct.
- The original owner of a large skull found in East Africa in 1959 was given the name *Nutcracker Man*, and acclaimed by *National Geographic* as evidence of man's evolutionary descent from the ape, but some time later even the palaeontologists who found it withdrew their extravagant claims, and it is now generally accepted that the skull was that of an extinct ape.
- In 1974, a tiny skeleton found in the Great Rift Valley, Ethiopia, was named *Lucy*, and said to date back at least three million years. The media took *Lucy* to their hearts, and she was widely promoted as having been the first ape to walk upright, but experts in the field have since reduced the so-called evidence to shreds.

There have been hundreds of other attempts to link *Homo sapiens* to apelike ancestors, but when exposed to honest investigation they have all met the same fate, and Darwin's idea that 'Man is descended from some less highly organized form,' remains pure guesswork. Fascinating drawings of apemen, with their barrel chests, jutting jaws and hairy legs, are the figments of human imagination. In the television première of *Ape Man: The Story of Human Evolution*, the high-profile American broadcaster Walter Cronkite told his audience, 'If you go back far enough, we and the chimps share a common ancestor. My father's father's father's father's father, going back maybe a half million generations — about five million years — was an ape.'[36] Cronkite earned a formidable reputation as a newscaster, but on this occasion he was parroting a prime-time fairy tale.

Phillip Johnson spells it out more fully: 'Instead of a fact we have a speculative hypothesis that says that living species evolved from ancestors which cannot be identified, by some much-disputed mechanism which cannot be demonstrated, and in such a manner that few traces of the process were left in the record *even though that record has been interpreted by persons strongly committed to proving evolution.*'[37] The last phrase makes Johnson's comments even more significant!

A 1980 *Newsweek* article confirmed that '… in the fossil record, missing links are the rule,' and went on to say, 'The more scientists have searched for the transitional forms that lie between species, the more they have been frustrated.'[38] As a result, scientists bent on establishing the evolutionary model have tried to do so by highlighting the similarities in the DNA of the various species. DNA (shorthand for deoxyribonucleic acid) is a molecule which includes long sequences of just four basic elements — adenine, guanine, cytosine and thymine, usually referred to as A, G, C and T. Yet, using these four 'letters' arranged in different ways, it spells out the blueprint for the production of all the proteins that an organism needs to grow and survive. In other words, it contains the genetic information necessary to form the chemicals and structures of life. Initial results in the field of so-called molecular evolution seemed to be getting somewhere, but this is proving to be another dead end. Apes and humans obviously have many physical similarities, and have similar DNA, but similarities have also been found among species where no close evolutionary link is claimed. Haemoglobin, the molecule that carries oxygen in red blood cells, is found in all vertebrates, including humans, but it also exists in earthworms, starfish, molluscs, in some insects and plants, and even in certain bacteria. Other evidence is equally confusing: when scientists examined the haemoglobin of crocodiles, vipers and chickens, they found that the crocodiles were more closely linked to the chickens than to their fellow reptiles.[39] In another test, an identical protein was found on the cell wall of both camels and nurse sharks.[40]

Similarity is not the same as relationship and, as far as evolution is concerned, molecular similarities produce more questions than answers. What is more, *Homo sapiens* is separated from all other species by characteristics that cannot be explained by even the smoothest transitions that evolutionists suggest. Here are some obvious examples:[41]

- We have vastly superior intelligence. We can accumulate, remember and evaluate masses of information on an immense variety of subjects, and then act rationally as a result.
- We possess qualities not only of consciousness, but of self-awareness, which lead us to reflect on our identity and relevance. We instinctively sense that we are neither atomic accidents nor educated apes. We think about meaning and purpose. We long for significance. We conceive aspirations and goals. We have an in-built sense of dignity.
- We are able to look beyond our own immediate and direct experiences. We think about death and what may lie beyond.
- We use propositional language, enabling us to write prose and poetry, translate languages, and to use words to express our emotions.
- We are capable of complex reasoning, lateral thinking and the development of theories and insights.
- We have mathematical skills, the ability to count to immense numbers, construct algebraic equations, and discuss issues in mathematics, statistics and science generally.
- We have an aesthetic dimension, enabling us to assess the relative qualities of form, texture, colour, order and design. We are endlessly creative in composing music and making pictures and other objects, not merely for survival but for our personal enjoyment.

- We have a moral dimension, a stubborn perception that there is a difference between right and wrong, together with a sense of ethical responsibility.
- We have a spiritual dimension, a sense that there *is* a world beyond the material.

These are just nine examples of the massive gulf between human beings and any other life form on our planet, and no evolutionary hypothesis has reduced it by so much as a millimetre. Ronald Nash, Professor of Philosophy at Western Kentucky University, makes the point precisely: 'Even if, for the sake of argument, we assume the truth of a universal evolutionary hypothesis, the fact that this process produced creatures with intelligence, creativity, self-consciousness and God-consciousness demands an explanation that Naturalism seems powerless to provide.'[42] The twentieth-century American politician William Jennings Bryan said the same thing in more down-to-earth terms: 'There is no more reason to believe that man descended from an inferior animal than there is to believe that a stately mansion has descended from a small cottage.'[43]

The first missing link

The only certain thing we can say about the missing links in the fossil record is that they are still missing — and the case against macro-evolution gets even stronger when we go one step further back. For the whole process of evolution to get started, there would first need to be one complete, self-contained, self-replicating life form. In ruling out God as the creator of life, evolutionists have turned to the so-called 'spontaneous generation' theory, which says that at some point in prehistory a random mixture of chemicals accidentally combined to produce the first living cell.

Darwin was so taken with this idea that he fantasized about 'some warm little pond, with all sorts of ammonia and phosphoric salts, light, heat, electricity, etc. present' and about 'a protein compound chemically formed ready to undergo still more complex changes'.[44]

The following statement on the Emmy-Award-winning PBS NOVA television programme *The Miracle of Life* is an example of the way in which one man's dream has become the media's dogma: 'Four and a half billion years ago, the young planet Earth was a mass of cosmic dust and particles. It was almost completely engulfed by the shallow primordial seas. Powerful winds gathered random molecules from the atmosphere. Some were deposited in the seas. Tides and currents swept the molecules together. And somewhere in this ancient ocean the miracle of life began... The first organized form of primitive life was a tiny protozoan (a one-celled animal). Millions of protozoa populated the ancient seas. These early organisms were completely self-sufficient in their seawater world. They moved about their aquatic environment feeding on bacteria and other organisms... From these one-celled organisms evolved all life on earth.'[45] Computer-generated graphics helped to make fascinating television, but one has only to list the phenomena taken for granted to realize that the material presented does not reflect science but fantasy. The cosmic dust and particles, the molecule-gathering winds and tides, the protozoa, the bacteria and the other organisms were all blithely assumed and pulled together to tell a politically correct story in which speculation masqueraded as information.

The fact is that, although countless experiments have tried to show that the first organisms could have arisen in this way from a sea of lifeless 'soup' on the early earth (biologist L. R. Croft says that at one stage 'A small cottage industry of primeval-soup workers was busily creating new concoctions'),[46] they have failed to provide evolution with the basis it desperately needs. At most, they have produced a few amino acids, which are almost infinitely less complex than the simplest protein molecules on which life depends. As Phillip Johnson points out, this is hardly surprising: 'There is no reason to believe that life has a tendency to emerge when the right chemicals are sloshing about in a soup. Although some components of living systems can be duplicated with very advanced techniques, scientists employing the full power of their intelligence cannot manufacture living organisms from amino acids, sugars, and the like. How then was the trick done before scientific intelligence was in existence?'[47] He may well ask!

When Darwin was fantasizing about his 'warm little pond', biologists knew little about biochemistry and even less about microbiology.

In Darwin's time, a biological cell was thought to be a simple 'blob' of disordered particles that needed little or no explanation, but we now know that there is no such thing as a primitive cell. Even *Mycoplasma genitalium*, the bacterium with the smallest known amount of genetic material, has 580,000 base pairs on its 482 genes.[48] DNA houses a staggering amount of genetic information. All the data needed to specify the design of a human being, including the arrangement of over 200 bones, 600 muscles, 10,000 auditory nerve fibres, two million optic nerve fibres, 100 billion brain-cell nerves and 400 billion feet of blood vessels and capillaries is packed into a unit weighing less than a few thousand-millionths of a gram, and several thousand million million times smaller than the smallest piece of functional machinery used by man. It has been said that on the same scale all the information needed to specify the design of every living species that has ever existed on our planet could be put in a teaspoon, with enough room left over for all the information in every book ever written.[49] Some 'blob'!

In *Evolution: A Theory in Crisis*, Michael Denton says that every living cell is 'an object of unparalleled complexity and adaptive design ... resembling an immense automated factory' and 'carrying out almost as many unique functions as all the manufacturing activities of man on earth'.[50] He goes on to say that this 'factory' has analogues of much of today's technology, including artificial languages and their decoding systems, memory banks for information storage and retrieval, elegant control systems regulating the automated assembly of parts and components, error fail-safe and proof-reading devices utilized for quality control, and assembly processes involving the principle of prefabrication and modular construction.[51] Yet evolutionists ask us to believe that the first living cells, 'far more complicated than any machine built by man and absolutely without parallel in the non-living world',[52] were not planned or designed, but came into being by sheer accident when winds and tides somehow stirred just the right molecules together in some kind of primordial soup.

What are the chances of this happening? Assuming the existence of all the necessary components — a huge assumption — Dr James Coppedge, an expert in the study of statistical probability, has calculated that the likelihood of a single protein molecule being arranged by chance is 1 in 10^{161}. As there are estimated to be only 10^{80} atoms in the entire universe, one would need 10^{81} universes for this to happen. This is, of course, ten times more universes than there are atoms in our own universe![53]

In the light of facts like these, many scientists have abandoned the idea that life could have arisen spontaneously from non-life. The British astrophysicist Sir Fred Hoyle called the odds against such a thing happening vast enough 'to bury Darwin and the whole theory of evolution'.[54] Elsewhere he wrote, 'The notion that … the operating programme of a living cell could be arrived at by chance in a primordial soup here on earth is evidently nonsense of a high order.'[55] This led him to embrace *panspermia*, the notion that the first living organisms on earth reached here from outer space, either accidentally or otherwise; but there is not a shred of evidence to back it up. Even if there were, it would take us no nearer an explanation of how life started.

Evolutionists usually insist that whatever the odds against the spontaneous generation of life, the sheer lapse of time gives an opportunity for anything to happen. In *The Chemistry of Life,* Harvard scientist George Wald, winner of the 1967 Nobel prize for physiology, claimed that this would make room for self-existent and self-organized matter: 'Time is the hero of the plot. Given enough time, anything can happen — the impossible becomes probable, the improbable becomes certain.'[56] Richard Dawkins suggests that 'the Replicator', a remarkable molecule with the ability to create copies of itself, came into being by accident at some point in the earth's prehistory. He admits that this is 'exceedingly improbable',[57] but in *The Blind Watchmaker* he covers himself by saying that '… given enough time, anything is possible.'[58] This is a popular ploy by evolutionists, but it embodies a fundamental fallacy — namely that because random changes go 'backward', as well as 'forward', no amount of time can, in and of itself, increase the possibility of anything coming into existence unless there are specific reasons for such a thing happening. The theory of evolution has none to offer.

After many years of study, and the publication of numerous research papers, Stephen Grocott, Fellow of the Royal Australian Chemical Institute, came to this conclusion: 'I am afraid that as a scientist I simply cannot say strongly enough that spontaneous origin of life is a chemical nonsense and, therefore, I am left with no alternative but to believe that life was created.'[59]

Square One

Darwin had little to say about the origin of life, and even less about the origin of matter, but any model relying on given phenomena can fairly be asked where the phenomena came from in the first place. We have already seen the impossible odds against life arising spontaneously from non-living material, but full-blown atheistic evolutionism can also be asked to explain the existence of such material in the first place. To be specific, where did the solids and liquids that made up Darwin's 'warm little pond' come from — let alone our planet and the vast physical universe beyond it? In his best seller *A Brief History of Time*, the British scientist Stephen Hawking calls Earth 'a medium-sized planet orbiting around an average star in the outer suburbs of an ordinary spiral galaxy'.[60] The 'average star' is the sun, which is one million times the size of Earth, while the 'ordinary spiral galaxy' is the Milky Way, which stretches some 621,000 million million miles across and contains about 100,000 million stars, yet is only one of about 100,000 million known galaxies. Evolution takes all of this for granted, but in doing so ducks a critical question: *where did this vast accumulation of matter come from?* There are only three possibilities.

1. Evolutionism's first choice is to say that *the entire universe is infinite and eternal.* This seems to be what the British philosopher Bertrand Russell

had in mind when he said, 'The world is simply there, and is inexplicable',[61] but while this kind of statement short-circuits the discussion, it does nothing to answer the question. In 1948 Fred Hoyle helped to popularize the so-called 'steady-state' theory, which maintained that the universe was infinite and eternal, and that as matter 'dies' through expansion it is replaced by other matter springing into existence. But, as philosopher William Lane Craig points out, the 'steady-state' model 'has been abandoned by virtually everyone'.[62]

2. The second possibility is that *the universe is self-created*, that it sprang into existence without any rhyme or reason. Versions of the spontaneous generation idea can be traced back nearly 3,000 years, and in the nineteenth century scientists began to speculate that it could account for the existence of the entire universe, but the whole idea flies in the face of logic. How could anything bring itself into existence unless it already existed in order to do so? The well-known statement *Ex nihilo, nihil fit* ('Out of nothing, nothing comes') is a fundamental axiom of natural law, and to believe that everything could come from nothing is to live in cloud cuckoo land. Peter Atkins, Professor of Physical Chemistry at Oxford University, makes a visit there whenever he insists that the entire universe is 'an elaborate and engaging rearrangement of nothing'.[63]

Even if we were to accept the popular notion of a 'Big Bang' occurring about fifteen billion years ago, we would still be left with no idea as to where the original 'singularity' came from, how time began, or what came before 'time zero'. Edgar Andrews, Emeritus Professor of Materials Science in the University of London, goes even further and reminds us that '… science, even at its most speculative, must stop short of offering any explanation or even description of the actual event of origin.'[64]

Eternality and spontaneous generation would obviously rule out creation by God, but they both come into fatal collision with two of the most fundamental laws of science known to man. These have to do with *energy* and *entropy*. In the context of these laws, everything in the natural world is one form of energy or another, while entropy is the measure of a system's lack of available energy to do or perform work. An intricate, complex and highly organized system is said to have low entropy, while a disorganized or 'run-down' system is said to have high entropy. The laws concerned are the First and Second Laws of Thermodynamics, which one writer has described as 'based upon more evidence and ... more universally applicable than any other principles in science'.[65]

The First Law, which has been called 'the most powerful and most fundamental generalization about the universe that scientists have ever been able to make',[66] categorically states that in any given system neither matter nor energy can be self-created or destroyed. One form of energy can be converted into another; matter can be converted into energy and energy into matter; but their sum total must remain the same. No new energy or matter is coming into being, nor is anything being annihilated. This rules out the contention that the formation of the natural world began spontaneously by natural processes.

The Second Law implies that over time any closed physical system becomes less ordered and more random — that is to say, entropy increases. Applied in simple terms, this means that our entire universe is like a clock that has been wound up and is now running down, becoming more and more disorganized as its energy becomes 'randomized'. This flatly contradicts biological evolution, which claims that chaotic and random changes lead to greater order and complexity. As Henry Morris says, 'The very idea of equating evolution with entropy is like equating east with west, or noon with midnight.'[67] The Second Law tells us that there is a universal tendency towards disintegration, decay and death; the theory of evolution claims exactly the opposite. It should not be too difficult to choose between a law that we can see at work all around us and a theory that proposes something that nobody has ever seen. The British physicist Sir Arthur Eddington, who dominated the world of stellar astronomy, had this advice for any evolutionist who tried to evade this particular issue: 'If your theory is found to be against the second law of thermodynamics I can give you no hope; there is nothing for it but to collapse in deepest humiliation.'[68] Evolutionists try to escape the implications of entropy by arguing that the Second Law applies only to closed systems, whereas the earth is not a closed system — in that it receives energy from the sun, and this energy can drive entropy backwards, towards greater order. But this is a false argument. No amount of random energy can create order unless the energy is harnessed by some pre-existing metabolic machinery. Simply put, only life can produce life.

3. This leaves just one alternative. As everything that had a beginning must have had a cause greater than itself, and the universe had a beginning, *it must therefore have had a transcendent, eternal and self-existent cause.* Referring to the First and Second Laws, Henry Morris comes to this conclusion: 'The only reasonable deduction from these scientific laws is that the world, with all its processes and with all its components, was brought into existence at some time in the past by means of creative and ordering processes which no longer exist and are therefore no longer available for scientific study. The First Law tells us either that the world has always existed in its present form or else that it was specially created at some time in the past. The Second Law tells us that it cannot always have existed in its present form or else it would already have completely disintegrated and died. The universe must therefore have had a beginning, and that beginning must have been by special creation.'[69]

All change?

Within about fifty years of Darwin's death in 1882, his 'general theory' was running out of steam. While biologists accepted that natural selection could preserve life by eliminating unfit elements, they realized that nowhere in *The Origin* is there a single case of its having produced evolutionary change leading to the creation of a new species. Experiments in plant biology consistently exposed a fundamental fallacy in Darwin's idea, and it became increasingly obvious that natural selection could never bring about organic evolution. A new line of approach was needed, and the answer seemed to lie in the field of genetics. What if genes underwent radical alterations (mutations)? Natural selection could then make use of these 'improved' genes and, given sufficient time, result in new and better species. This 'synthetic theory' soon became all the rage and, under the general title of neo-Darwinism, it has now so completely replaced the original model that Sylvia Baker, a respected writer in the field, claims, 'The modern theory of evolution … stands or falls on this question of mutation.'[70] Everybody agrees that both natural selection and genetic changes take place, but there are at least four compelling reasons for discarding the idea that they combine to produce the new life forms that evolution demands.

Firstly, *mutations* (other than those artificially induced in a laboratory) *occur extremely rarely*, something like once in every ten million duplications of a DNA molecule.[71] After studying mutations in many generations of bacteria, Pierre Grassé found that no essential changes had developed. As bacteria multiply 400,000 times faster than humans, Grassé's findings equate to millions of years within the human species. Dismissing the possibility of 'thousands and thousands of lucky, appropriate events' occurring to produce the mutational changes on which neo-Darwinianism leans, Grassé wrote, 'There is no law against daydreaming, but science must not indulge in it.'[72]

Secondly, far from producing strong, improved genes that would promote evolution, *virtually all mutations* (999 out of every 1,000) *are harmful*, weakening the organism or destroying it altogether. Likening a genetic mutation to a typing error, the Canadian medical professor Magnus Verbrugge writes, 'Typing errors rarely improve the quality of a written message; if too many occur, they may even destroy the information contained in it.'[73]

Phillip Johnson is even more graphic in dismissing the possibility of a genetic mutation's contributing to even one component of an improved species: 'To suppose that such a random event could reconstruct even a single complex organ like a liver or kidney is about as reasonable as to suppose that an improved watch can be designed by throwing an old one against a wall.'[74]

Thirdly, while beneficial mutations leading to improved species would entail a massive increase in the genome's information, *no such increase has ever been observed*. Biophysicist Lee Spetner, who taught information and communication theory at America's Johns Hopkins University, underlines the significance of this fact: 'The neo-Darwinians would like us to believe that large evolutionary changes result from a series of small events if there are enough of them. But if these events all *lose* information they can't be the steps in the kind of evolution [the theory] is supposed to explain, no matter how many mutations there are. *Whoever thinks macro-evolution can be made by mutations that lose information is like the merchant who lost a little money on every sale but thought he could make it up on volume...* The failure to observe even one mutation that adds information is more than just a failure to find support for the theory. *It is evidence against the theory.*'[75]

Fourthly, for any new, functional organ to be effective it would have to arrive on the scene all at once, as a complete, operating entity — but evolutionists say that mutation takes place in microscopic increments, each of which achieves almost nothing in and of itself. If this is the case, how can this kind of process give us the finished product? Take the human eye, for example, which comes with automatic aiming, focusing and aperture adjustment and has 130 million receptor cells, 124 million of which are rod-shaped and differentiate between light and darkness, and six million of which are cone-shaped and can identify up to eight million variations of colour. Are we seriously to believe that this staggeringly sophisticated organ came into being through a step-by-step, accidental, purposeless, trial-and-error process that took millions of years? Some evolutionists stubbornly maintain that this is the case, and that even at an early stage of the process, one per cent of an eye would have been better than no eye at all. But this is to make the elementary mistake of confusing one per cent of an eye with one per cent of normal vision, which is a very different matter. One per cent of an eye would give no vision at all. What is more, even if every one of the eye's components were in place, they would achieve nothing unless they were precisely 'wired' to an amazing complex of millions of nerve cells in the brain and elsewhere in the body. Must we also assume that all these millions of other related features necessary before the eye could function also developed simultaneously and in the same random way?

The complexity of the eye raises another enormous problem for the evolutionist. Darwin said, 'If it could be demonstrated that any complex organ existed which could not possibly have been formed by numerous, successive, slight modifications, *my theory would absolutely break down.*'[76] As the American professor of biochemistry Michael Behe brilliantly demonstrates in his superb book *Darwin's Black Box,*[77] we now know that the eye is just one of many such systems in the human body alone. Where does that leave Darwin's challenge?

Fifthly, no plant or animal lives long enough to allow the millions of micro-mutations that would be needed to transform it into a different, 'improved' species. Magnus Verbrugge gives this illustration: 'How likely is it that random mutations will come together and co-ordinate to form just one new structure? Let's say the formation of an insect wing requires only five genes (a very low estimate)... The probability of two non-harmful mutations occurring [simultaneously] is one in one thousand million million. *For all practical purposes, there is no chance that all five mutations will occur within the life cycle of a single organism.*'[78] This alone is enough to torpedo the 'synthetic theory', but it sinks without trace when we realize that a single organism is made up of many structures *that must appear at the same time, working together as an integrated whole.* Even a convinced evolutionist like the American zoologist George Gaylord Simpson admitted that if there was an effective breeding population of 100 million individuals, and they could produce a new generation every single day, the likelihood of producing good evolutionary results from mutations could be expected only about once in 247 billion years — far longer than even the most extravagant estimation of the age of the earth.[79]

Although neo-Darwinism is widely accepted as being beyond dispute, its credibility evaporates as soon as we expose it to honest scientific analysis. Natural selection destroys unfit organisms, mutations result in a loss of genetic information, and time inevitably leads to decay and death. Changing the word 'ant' into 'antelope' can be done in a second or two by adding five letters; changing an ant into an antelope by a succession of rare, harmful and completely random steps is another matter! Sir Ernst Chain, co-holder of a 1945 Nobel Prize for his work in the development and use of penicillin, was not exaggerating when he said that the development and survival of the fittest by chance mutations was 'a hypothesis based on no evidence and irreconcilable with the facts'.[80]

The meaning behind the myth

The case for macro-evolution is so persistently and passionately advocated that millions of people with no scientific expertise assume it must be true. Any statement beginning with the claim that scientists have 'proved' or 'shown' evolution to be a fact tends to be taken at its face value, as there seems to be no point in arguing with the experts. Yet there are at least two reasons why we should resist this response.

The first is that science is not a finished product, but an ongoing search for truth, a process of learning in which from time to time things previously said to be true are found to be false. In a *New Scientist* cartoon, a student responds to a teacher's statement by asking, 'Are you sure that is the right answer?' When the teacher confirms that it is, the student replies, 'But you told us the exact opposite yesterday!' 'Yes, I did,' the teacher tells her, 'but that was yesterday. We must remember that science is making tremendous strides'! We need to realize that in true science, the latest word is not necessarily the last word. Sir Karl Popper, arguably the world's best-known philosopher of science, even goes so far as to say, 'Every scientific statement must remain tentative for ever.'[81]

The second is that there are many things beyond the reach of science. It can never explain why the world came into being, where energy and matter came from, why there are consistent and dependable natural laws — nor can it say anything definitive about the theory of evolution. As even a convinced evolutionist like George Gaylord Simpson confirms, 'It is inherent in any definition of science that statements that cannot be checked by observation are not really about anything ... or at the very least they are not science.'[82] For any idea to qualify as a credible scientific theory, it must be backed up by events, processes or properties that can be observed and tested. Scientists usually add another qualification —

that it must be possible to set up an experiment, the failure of which would prove the theory wrong. The theory of evolution obviously meets none of these criteria. No human witnesses saw the world come into existence — no one saw the first spark of life, a fish turning into an amphibian, an amphibian into a reptile, a reptile into a bird or a mammal. In fact, *there is no first-hand evidence that a single species ever came into being by natural processes.* Nor can the theory of evolution be proved to be true by the failure of an experiment set up for the purpose because, as an article in *Nature* makes clear, 'No one can think of ways in which to test it.'[83] The whole molecules-to-man story is just that — *a story.* It is passionately told, vehemently argued and decorated with thousands of fascinating 'exhibits', but it has no reliable scientific basis.

This raises the obvious question: why do the storytellers work so hard to convince us that they are telling the truth? The answer is not difficult to find, and Richard Dawkins gave a significant clue during the BBC television programme *Soul of Britain*.[84] Asked about his reaction to the fact that most people in Britain believe in God, he replied, 'I am unhappy to be living in a society where I think the majority of people are deluded. I'd love to do something about it, which is why I write the books I do.' This makes it clear that although Dawkins writes about biology, zoology and the like, he has a hidden agenda, as he confirmed by adding, 'I devoutly wish that we did live in a post-God society.'[85]

Two years after *The Origin* was published, Adam Sedgwick, Woodwardian Professor of Geology at Cambridge University, wrote, 'From first to last it is a dish of materialism cleverly cooked up... And why is this done? For no other reason, I am sure, except to make us independent of a Creator.'[86]

Julian Huxley confirmed this at a 1959 conference to mark the centenary of the publication of *The Origin*: 'Darwin's real achievement was to remove the whole idea of God as the Creator of organisms from the sphere of rational discussion.'[87]

Since then a succession of high-profile scientists such as Nobel Prize winner Harold Urey have openly admitted that their commitment to materialism (in other words to a world-view that begins by excluding

God) has led them to embrace the theory of evolution not as a scientific conclusion, but as an act of faith: 'All of us who study the origin of life find that the more we look into it, the more we feel it is too complex to have evolved anywhere. We all believe *as an article of faith* that life evolved from dead matter on this planet.'[88] Harvard's George Wald is even more blatant: 'When it comes to the origin of life on this earth, there are only two possibilities: creation or spontaneous generation. There is no third way. Spontaneous generation was disproved 100 years ago, but that leads us to only one other conclusion, that of supernatural creation. We cannot accept that on philosophical grounds; *therefore we choose to believe the impossible, that life arose spontaneously by chance.*'[89]

These examples confirm that, in the absence of any factual basis, evolution is a belief-system, which in turn means that *evolutionism is not a science but a religion.* In the introduction to the 1971 reprint of *The Origin*, the British biologist Harrison Matthews wrote that belief in the theory of evolution was 'exactly parallel to belief in special creation' and that evolution was 'simply a satisfactory faith on which to base our interpretation of nature'.[90] With particular reference to the question of fossils, L. T. More said much the same thing: 'The more one studies palaeontology, the more certain one becomes that evolution is based on faith alone; exactly the faith which it is necessary to have when one encounters the great mysteries of religion.'[91] This is very different from the popular idea that 'molecules-to-man' evolution is solidly based on science!

Consequences

Accepting the theory of evolution is much more than an intellectual option. It means pinning one's thinking to a materialistic belief-system that reduces everything to nature and natural processes — and this has far-reaching consequences. William Provine, Professor of History and Biological Sciences at Cornell University, who calls himself 'a total atheist', gives us more than an inkling of where it leads: 'Let me summarize my views on what modern evolutionary biology tells us loud and clear…

There are no gods, no purposes, no goal-directing forces of any kind. There is no life after death... There is no ultimate foundation for ethics, no meaning to life, and no free will for humans, either.'[92] This makes it clear that evolution is not some kind of philosophical toy we can bring out occasionally for our interest or amusement. It radically affects every part of life. If the world is 'just there', if life is the result of a fascinating fluke, and if we ourselves are nothing more than biological accidents, we are faced with an avalanche of questions:

- If information came into being by chance, how can we know that anything is true?
- How can evolution account for the universal and invariable laws of logic, on which all our thinking depends? On what basis can we study the world in any coherent way and come to sensible conclusions about it?
- If the brain is nothing more than an accident of biological evolution, why should we trust its ability to tell us so? How can chance accumulations of atoms and molecules decide that that is what they are?
- How can shrink-wrapped bags of biological elements have any rights to justice, freedom, possessions or happiness — or even to life itself? What gives us any greater value than rocks or reptiles, trees or termites?
- How did the products of a succession of genetic flukes learn to remember the past, evaluate the present and wonder about the future?
- If we are what someone called 'computers made of meat', how did we acquire an aesthetic dimension, enabling us to appreciate beauty in nature and art, when doing this makes no contribution to evolution?
- Why should we look for purpose or meaning in life? What is the sense in genetically programmed machines talking about 'quality of life' and 'values', or concerning themselves with aims or aspirations?

- As it is impossible to jump from atoms to ethics and from molecules to morality, why do we have an inbuilt sense of right and wrong? Where did conscience come from, and why does it have such amazing power? Why do we sometimes feel guilty or ashamed?
- Why do we have a sense of obligation or responsibility to other people? Why should mere blobs of animate matter be concerned for the temporary well-being of other blobs if both are on their way to extinction?
- How can we live — or die — with dignity if our existence is meaningless? Why do we take ourselves so seriously if Richard Dawkins is right to say that we are nothing but 'jumped-up apes'?[93]
- If the survival of the fittest is evolution's greatest prize, why should we not encourage the non-survival of the unfit? Why should we care for the frail, the mentally defective, the chronically sick, the senile, or the starving? Should we not give evolution a helping hand by getting rid of them — and the sooner the better?
- If humans are no more than grown-up germs, why are we the only species preoccupied with death? Why should approaching it — or delaying it — be of the slightest concern to us? If we began as a fluke, live out a farce and end as fertilizer, what hope or help can we give to someone who is dying?
- Why is our sense of spirituality so strong that man has been called 'a religious animal'? Is this something we should expect to happen to dust left around for millions of years?

The French philosopher Jean-Paul Sartre vehemently denied the existence of God, but towards the end of his life admitted that atheism was 'a cruel, long-term business'.[94] He could have said the same about the theory of evolution. Claiming to believe in it is one thing, but consistently living with all its implications is another matter. Can we seriously and consistently live as if we owed our existence to millions of mindless accidents, as if there was no rational or moral order, and as if we were ultimately just manure in the making?

There is an alternative…

Not evolution — revolution!

Any world-view worth holding must take in the past, the present and the future. To put this more directly, it must address questions about our origin, our everyday life and our destiny. One that does all of these things is to be found in the world's most reliable database — the Bible. Those who reject this assessment of the Bible have almost certainly never read it with an open mind. Not only is its literary integrity far greater than that of any other literature known to man, but archaeology, astronomy, geology and biology have all endorsed it, and no scientific discovery has ever contradicted anything it says. As great an authority as Sir Isaac Newton, universally recognized as the father of modern science, called it 'a rock from which all the hammers of criticism have never chipped a single fragment'. This is exactly what we should expect if the Bible is what it claims to be — 'the living and enduring word of God'.[95]

The Bible relates to the past

The origin of the universe required time, intelligence, energy, space and matter — and all five are to be found in the Bible's first ten words: 'In the beginning God created the heavens and the earth.'[96] In the original Hebrew, the phrase translated 'the heavens and the earth' means everything that exists outside of God himself — and 'everything' means exactly what it says, from galaxies to grains of sand, angels to asteroids, tigers to time, space to spiders, and light to the laws of physics. The Bible could hardly be more specific: it refers to 'the living God, who made heaven and earth and sea and everything in them'.[97] It does not give us a detailed explanation of *how* creation took place, nor does it tell us precisely *when* it happened. Its main concern is to make it clear that the universe, including time, space and the laws of nature, came into existence by divine fiat. God did not create because he had to, but because he chose to, and nothing in creation needs any justification beyond the fact that God willed it and brought it into being. What other reason could we demand when God is 'exalted as head over all'[98] and 'Wisdom and power are his'?[99] In *A Brief History of Time*, Stephen Hawking said that if we were to discover a comprehensive theory (the so-called 'Theory of Everything') explaining why the universe is exactly the way it is, '... it would be the ultimate triumph of human reason — *for then we would know the mind of God*.'[100]

The universe exists to reflect the glory of its Maker, whose free, independent and sovereign will is the originating cause of all things. This explains why the laws of science are consistent throughout all of time and space. With no religious axe to grind, physicist Paul Davies, Professor of Natural Philosophy at the University of Adelaide, says, 'It is hard to resist the impression that the present structure of the universe, apparently so sensitive to minor alterations in the numbers, has been rather carefully thought out.'[101] The British scientist Sir John Houghton goes further: 'The order and consistency we see in our science can be seen as reflecting orderliness and consistency in the character of God himself.'[102] Does it take any more faith to believe this than to believe that the amazing order, harmony and beauty we see in the natural world is a gigantic fluke, that life itself sprang into existence by chance, that logic is sheer luck and that the vast amount of information in living things had no intelligent source? The universally respected scholar and author C. S. Lewis came to this conclusion: 'No philosophical theory which I have yet come across is a radical improvement on the words of Genesis, that "In the beginning God made heaven and earth".'[103]

The Bible relates to the present

Towards the end of its record of creation, the Bible tells us:

> So God created man
>   in his own image,
> in the image of God
>   he created him,
> male and female
>   he created them. [104]

This simple statement is profoundly relevant to our everyday lives. It explains precisely and uniquely why we can properly claim to have greater dignity than donkeys or dandelions. The distinguished modern scholar Francis Schaeffer hits the nail on the head: 'The Bible tells me who I am... Suddenly I have value, and I understand how it is that I am different... A man is of great value not for some less basic reason *but because of his origin.*'[105]

Our creation by the one who is 'majestic in holiness'[106] explains why we have a sense of right and wrong. The Bible says even of those who reject God that the requirements of his holy law 'are written on their hearts'.[107] The conscience is 'God's calling card',[108] an inescapable reminder that we have a moral obligation to our Maker. This is why everybody whose thinking is straight *knows* that dishonesty, immorality, envy, greed and selfishness are wrong, and why their opposites are right.

Our creation by God relates directly to the value we place on other people. As Francis Schaeffer says, 'If man is not made in the image of God, then nothing stands in the way of inhumanity... Human life is cheapened.'[109] An evolutionary world-view reduces us to atoms and molecules, leaving us free to treat each other in any way we choose in order to get what we want, from abortion on demand to euthanasia. The Bible says exactly the opposite, and tells us that human life is sacred because it is sanctified by God and that, whatever their present weakness, inability or defects, we should treat others on the basis that they 'have been made in God's likeness'.[110]

Yet the God revealed in the Bible is not some kind of cosmic clockmaker who 'wound the world up' and now takes no interest in it. He not only 'gives all men life and breath and everything else',[111] he 'has compassion on all he has made'.[112] What is more, to those who trust him he proves to be a 'refuge and strength, an ever-present help in trouble'.[113] It is the testimony of millions of people over thousands of years that in times of pain, sorrow, loneliness, guilt, depression and confusion they have found God to be a dynamic, liberating reality, bringing joy and peace where neither seemed possible.

The Bible relates to the future

It not only confirms that we are 'destined to die', [114] it says that God has 'set eternity in the hearts of men'.[115] We have an inbuilt awareness that death is *not* the end, and that beyond this mortal life lies immortality. Many people with no more than a vague belief in him assume that as 'God is love'[116] they will automatically go to heaven and enjoy his presence for ever. But this ignores the Bible's teaching that 'Nothing impure will ever enter [heaven]'[117] and that, as 'the righteous Judge',[118] God will condemn to 'eternal punishment'[119] all those whose sins are not forgiven.

As 'All have sinned and fall short of the glory of God',[120] our prospects would seem to be truly terrible — but God has intervened! The Bible's central message is that in the person of his Son, Jesus Christ, he 'came into the world to save sinners'.[121] In his life, Jesus gave us a perfect example of how we should live. In his death, he paid the penalty we deserve: 'Christ died for sins once for all, the righteous for the unrighteous, to bring you to God.'[122] In his resurrection from the dead not only was he 'declared with power to be the Son of God',[123] but he provided the means by which all those who put their trust in him can be certain that their sins are forgiven and that by the grace of God they will spend eternity in his glorious, sinless, deathless, endless presence.

The Word of God contradicts the theory of biological evolution, but promises spiritual revolution. Why settle for speculation when you can experience the reality of God's presence and power in your life?

Notes

1. *The Concise Oxford Dictionary,* Oxford, 7th edition, 1988, p.334.
2. Cited by John D Currid, 'From the Renaissance to the age of Naturalism', in *Building a Christian World View,* ed. W. Andrew Hoffecker, Presbyterian & Reformed Publishing Company, vol. 1, pp.154-5 (emphasis added).
3. *Ibid.* (emphasis added).
4. Theodosius Dobzhansky, 'Changing Man', in *Science* 155: 409.
5. Cited by John Wright, *Designer Universe,* Monarch Publications, p.61.
6. Michael Denton, *Evolution: A Theory in Crisis,* Adler & Adler, p.15.
7. H. J. Muller, 'Is Biological Evolution a Principle of Nature that has been well established by Science?' (Privately duplicated and distributed by author, May 1966).
8. Carl Sagan, *Cosmos,* Random House, p.27.
9. Cited by Phillip E. Johnson, *Darwin on Trial,* Monarch Publications, p.9.
10. Julian Huxley, 'Issues in Evolution', in *Evolution after Darwin,* vol. 3, ed. Sol Tax, Chicago University Press.
11. Cited by Malcolm Bowden, *The Rise of the Evolution Fraud,* Sovereign Publications, p.218.
12. Cited *ibid.,* p.216.
13. Personal letter to Luther D. Sutherland, cited by Sutherland in *Darwin's Enigma,* Master Book Publishers, p.89.
14. Henry M. Morris, *The Long War Against God,* Baker Book House, p.156.
15. Pierre Grassé, *Evolution of Living Organisms,* Academic Press, p.4 (emphasis added).
16. See Johnson, *Darwin on Trial,* p.175.
17. Charles Darwin, *The Origin of Species,* J. M. Dent & Sons Ltd, pp.292-3.
18. David M. Raup, 'Conflicts between Darwin and Palaeontology', in *Field Museum of Natural History Bulletin,* vol. 50, p.25 (emphasis added).
19. Cited by Scott M. Huse, *The Collapse of Evolution,* Baker Books, p.58 (emphasis added).
20. Marvin Lubenow, *Bones of Contention,* Baker Books, p.182.
21. David Kitts, *Evolution* 28: 467 (emphasis added).
22. Cornelius G. Hunter, *Darwin's God,* Brazos Press, p.69.
23. Cited by Johnson, *Darwin on Trial,* p.54.

24. Johnson, *Darwin on Trial*, p.54.
25. Colin Patterson, *The Listener* 106: 390.
26. Stephen Jay Gould, 'Evolution's erratic pace', *Natural History,* vol. LXXXVI(5), p.14 (emphasis added).
27. Colin Patterson, *Evolution,* British Museum of Natural History, p.133.
28. Jeremy Rifkin, *Algeny,* Viking Press, p.188.
29. Stephen J. Gould, *The Panda's Thumb,* W. W. Norton & Co., p.184.
30. Cited by Johnson, *Darwin on Trial,* p.59 (emphasis added).
31. Patterson, *The Listener* 106: 390 (emphasis added).
32. Richard Lewontin, *New York Review of Books,* 9 January 1997 (emphasis added).
33. Cited by A. J. Monty White, *Wonderfully Made,* Evangelical Press, p.33 (emphasis added).
34. Lubenow, *Bones of Contention,* p.65.
35. Ian T. Taylor, *In the Minds of Men,* TFE Publishing, p.227.
36. *Ape Man: The Story of Human Evolution,* Arts and Entertainment Network, 4 September 1994.
37. Phillip Johnson, privately circulated article cited by Malcolm Bowden, *Science vs. Evolution,* Sovereign Publications, p.227 (emphasis added).
38. *Newsweek,* 3 November 1980.
39. See Henry M. Morris and Gary E. Parker, *What is Creation Science?,* Master Books, pp.52-61.
40. Cited in *New Scientist* 160 (2154): 23.
41. I have dealt with this more fully in *Does God Believe in Atheists?,* Evangelical Press, pp.324-31.
42. Ronald Nash, *Faith and Reason,* Zondervan Publishing House, p.138.
43. Cited by Edyth Draper, *Draper's Book of Quotations for the Christian World,* Tyndale House Publishers, p.111.
44. Cited by Johnson, *Darwin on Trial,* p.101.
45. NOVA: *The Miracle of Life,* WGBH Educational Foundation.
46. L. R. Croft, *How Life Began,* Evangelical Press, p.42.
47. Johnson, *Darwin on Trial,* p.103.
48. C. M. Fraser *et. al.,* 'The Minimal Gene Complement of *Mycoplasma genitalium*', in *Science* 270 (5235): 397-403.
49. See T. S. Kuhn, *The Structure of Scientific Resolutions,* 2nd edition, University of Chicago Press, p.69.
50. Denton, *Evolution: a Theory in Crisis,* p.329.
51. *Ibid.*
52. *Ibid.,* p.250.
53. See James F. Coppedge, *Evolution: Possible or Impossible?,* Probability Research on Molecular Biology, pp.110ff.
54. Fred Hoyle, 'Hoyle on Evolution', *Nature* 294: 148.
55. Fred Hoyle, 'The big bang in astronomy', *New Scientist* 92: 521.

56. George Wald, 'The Origin of Life', in *The Physics and Chemistry of Life*, Simon and Schuster, p.12.
57. Richard Dawkins, *The Selfish Gene*, Oxford University Press, p.16.
58. Richard Dawkins, *The Blind Watchmaker*, W. W. Norton & Co., p.139.
59. Stephen Grocott, *In Six Days*, ed. John F. Ashton, p.136.
60. Stephen Hawking, *A Brief History of Time*, 1995 edition, Bantam Books, pp.139-40.
61. Cited by Gary Scott Smith, 'Naturalist Humanism', in *Building a Christian World View*, vol. 1, p.174.
62. William Lane Craig, *Reasonable Faith*, Crossway Books, p.103.
63. *Daily Telegraph*, 6 April 1998.
64. Edgar Andrews, *God, Science and Evolution*, Evangelical Press, p.35.
65. Henry M. Morris, *Evolution and the Modern Christian*, Presbyterian and Reformed Publishing Company, p.45.
66. Isaac Asimov, 'In the Game of Entropy and Thermodynamics You Can't Even Break Even', in *Journal of the Smithsonian Institute*, June 1990, p.6.
67. Henry M. Morris, *The God Who is Real*, Baker Book House, p.46.
68. Arthur S. Eddington, *The Nature of the Physical Universe*, Macmillan, p.74.
69. Morris, *Evolution and the Modern Christian*, p.48.
70. Sylvia Baker, *Bone of Contention*, Evangelical Press, p.19.
71. Gary E. Parker, *Creation — The Facts of Life*, Creation-Life Publishers, p.63.
72. Pierre Grassé, *Traité de zoologie*, Tome VIII, Masson.
73. Magnus Verbrugge, *Alive: An Enquiry into the Origin and Meaning of Life*, Ross House Books, p.12.
74. Johnson, *Darwin on Trial*, p.37.
75. Lee Spetner, *Not by Chance!*, The Judaica Press, p.160 (emphasis added).
76. Darwin, *The Origin of Species*, 8th edition, p.154 (emphasis added).
77. Michael Behe, *Darwin's Black Box — The Bio-Chemical Challenge to Evolution*, The Free Press.
78. Verbrugge, *Alive: An Enquiry into the Origin and Meaning of Life*, p.13 (emphasis added).
79. Cited by Nicholas Comninellis, *Creative Defense*, Master Books, p.81.
80. Cited by Edward F. Block, *Special Creation vs. Evolution*, Southwest Bible Church, p.5.
81. Karl Popper, *The Logic of Scientific Discovery*, Unwin Hyman Ltd, p.28.
82. George Gaylord Simpson, *Science* 143: 769.
83. L. C. Birch and P. R. Elrich, *Nature* 214: 239.
84. *Soul of Britain*, BBC2, 11 June 2000.
85. *Ibid.*
86. Cited by R. Clark, *The Survival of Charles Darwin*, Random House, p.139.
87. Cited by Don Batten, *In Six Days*, p.354.
88. *Christian Science Monitor*, 4 January 1962 (emphasis added).
89. Cited by Huse, *The Collapse of Evolution*, p.3 (emphasis added).

90. Cited by Taylor, *In the Minds of Men*, p.394.
91. Cited by R. L. Wysong, *The Creation-Evolution Controversy*, Inter-Varsity Press, p.31.
92. William B. Provine, *Origins Research* 16 (1/2): 9.
93. *Sunday Telegraph*, 18 October 1998.
94. Jean-Paul Sartre, *Works*, Penguin Books, p.65.
95. 1 Peter 1:23.
96. Genesis 1:1.
97. Acts 14:15.
98. 1 Chronicles 29:11.
99. Daniel 2:20.
100. Hawking, *A Brief History of Time*, p.193 (emphasis added).
101. Paul Davies, *God and the New Physics*, Touchstone Books, p.189.
102. John Houghton in *Real Science, Real Faith*, ed. R. J. Berry, Monarch Publications, p.46.
103. C. S. Lewis, *Miracles*, Collins, p.37.
104. Genesis 1:27.
105. Francis Schaeffer, *Genesis in Time and Space*, Hodder & Stoughton, p.51 (emphasis added).
106. Exodus 15:11.
107. Romans 2:15.
108. C. Stephen Evans, *The Quest for Faith*, Inter-Varsity Press, p.48.
109. Schaeffer, *Genesis in Time and Space*, p.27.
110. James 3:9.
111. Acts 17:25.
112. Psalm 145:9.
113. Psalm 46:1.
114. Hebrews 9:27.
115. Ecclesiastes 3:11.
116. 1 John 4:8.
117. Revelation 21:27.
118. 2 Timothy 4:8.
119. Matthew 25:46.
120. Romans 3:23.
121. 1 Timothy 1:15.
122. 1 Peter 3:18.
123. Romans 1:4.

In *Does God believe in Atheists?* John Blanchard deals in depth with many of the issues raised in this booklet. The book may be ordered direct from the publisher at one of the following addresses:

Evangelical Press, Faverdale North, Darlington DL3 0PH, England

Evangelical Press USA, P. O. Box 84, Auburn, MA 01501, USA

e-mail: sales@evangelicalpress.org

If, as a result of reading of this book, you would like to learn more about the Bible's teaching, you are invited to write to John Blanchard at the publisher's address for a free copy of either his booklet *Ultimate Questions*, which is a brief introduction to the Christian faith, or, if you have already come to trust in God, *Read Mark Learn,* his book of guidelines for personal Bible study.